DISCARD

j570 Terp, Gail
TER
     Proboscis monkeys

05-11-23   CHILDREN

Curious Creatures

PROBOSCIS MONKEYS

GAIL TERP

Black Rabbit Books

Bolt is published by Black Rabbit Books
P.O. Box 227, Mankato, Minnesota, 56002
www.blackrabbitbooks.com
Copyright © 2023 Black Rabbit Books

Marysa Storm, editor; Michael Sellner, designer and photo researcher

All rights reserved. No part of this book may be reproduced, stored in a retrieval system or transmitted in any form or by any means, electronic, mechanical, photocopying, recording, or otherwise, without written permission from the publisher.

Library of Congress Cataloging-in-Publication Data
Names: Terp, Gail, 1951- author.
Title: Proboscis monkeys / by Gail Terp.
Description: Mankato, Minnesota : Black Rabbit Books, [2023] | Series: Bolt. Curious creatures | Includes bibliographical references and index. | Audience: Ages 8-12 | Audience: Grades 4-6 | Summary: "Proboscis monkeys are animals unlike any other, and curious and reluctant readers alike will be captivated by these exceptional creatures' life cycle, habitats, diet, and threats to survival through carefully leveled text, detailed infographics that emphasize visual literacy, and engaging, colorful photography"– Provided by publisher.
Identifiers: LCCN 2020016583 (print) | LCCN 2020016584 (ebook) | ISBN 9781623105679 (hardcover) | ISBN 9781644664926 (paperback) | ISBN 9781623105730 (ebook)
Subjects: LCSH: Proboscis monkey–Juvenile literature.
Classification: LCC QL737.P93 T47 2023 (print) | LCC QL737.P93 (ebook) | DDC 599.8/6-dc23
LC record available at https://lccn.loc.gov/2020016583
LC ebook record available at https://lccn.loc.gov/2020016584

Image Credits

AgeFotostock: Anup Shah/AnimalsAnimals, 10–11; Alamy: Fredrik Stenström, 8–9; JUAN CARLOS MUÑOZ, 1; Michael S. Nolan, 27; Rosanne Tackaberry, 20–21; Dreamstime: Lukas Zeman, 12–13; Mario Madrona Barrera, 18–19; Sergey Uryadnikov, 16–17; Getty: ade ibadurrohman, 4–5; Ger Bosma, 26–27; Sylvain CORDIER, 22–23; Minden: Anup Shah, 22–23; Suzi Eszterhas, 3, 20–21, 31; Sylvain Cordier, 10–11; Shutterstock: Berendje Fotografie, Cover; irin-k, 25; Kozyreva Elena, 18–19; kenary820, 32; mastapiece, 14–15; Mazur Travel, 12–13; Michal Sloviak, 25; natika_art, 18–19, 28; OlegD, 6–7; Praisaeng, 25; Ruslan Gubaidullin, 6–7, 25; Rich Carey, 26; Sergey Uryadnikov, 12–13; Taweesak Sriwannawit, 25; wasanajai, 25; Yusnizam Yusof, 10–11; Wikimedia Commons, fvanrenterghem, 25

Contents

CHAPTER 1
A Proboscis Monkey in Action............4

CHAPTER 2
Food and Home......10

CHAPTER 3
The Family..........16

CHAPTER 4
Danger!..............24

Other Resources...........30

CHAPTER 1

A Proboscis Monkey in Action

A male proboscis monkey sits in a tree, keeping watch. Several females stay close, grooming their young. The male spots a leopard. Leopards are never good news. It's time to warn off this **predator**. The male's huge nose grows even bigger. The monkey honks. To add to the noise, it shakes the tree branches. The females begin to screech. The leopard leaves to find a meal somewhere else.

How Big Are Proboscis Monkeys?

**HEAD AND BODY LENGTH
24 to 30 INCHES**
(61 to 76 centimeters)

A Big Nose

All of these monkeys have big noses. But only the males grow giant noses called proboscises. These large, flexible noses can be up to 7 inches (18 cm) long. A long nose seems to **attract** females. The bigger the nose, the more females in a male's group.

PARTS OF A PROBOSCIS MONKEY

- NOSE
- EYES

CHAPTER 2

and Home

Proboscis monkeys spend their days resting, traveling, and eating. These monkeys sometimes feed on bugs. But they mostly eat leaves and unripe fruit. These foods can be hard to **digest**. But the monkeys' big stomachs are full of **bacteria**. The bacteria helps them break down their food.

Males with larger noses can honk louder.

Proboscis Monkey Habitats

MANGROVE FORESTS

RAIN FORESTS

PALM SWAMPS

Where They Live

Proboscis monkeys only live on the Asian island Borneo. They live near water and spend most of their time in trees.

These monkeys are good swimmers. Their webbed feet help them move easily through water. They often jump into the water from trees.

WHERE PROBOSCIS MONKEYS LIVE

Proboscis Monkey Range Map

CHAPTER 3

The

Proboscis monkeys are **social**. They often live in groups. For safety, groups come together at night to sleep. Large groups are harder for predators to attack than smaller groups. The monkeys work together to scare away attackers. They break into their smaller groups in the morning.

Groups won't sleep in the same trees two nights in a row.

COMPARING WEIGHTS

Giving Birth

Males **mate** with the females in their groups. About five months later, females give birth to one baby. They give birth while sitting in trees. Babies have blue faces. By about eight months old, their faces will be pink like their parents.

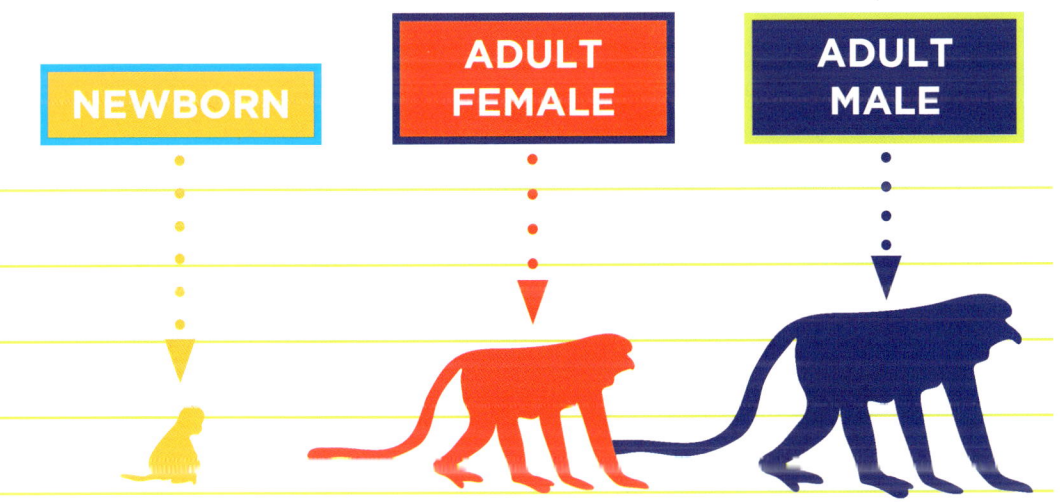

NEWBORN — about 1 POUND (0.45 kg)

ADULT FEMALE — up to 22 POUNDS (10 kg)

ADULT MALE — up to 50 POUNDS (22.7 kg)

Growing Up

Young proboscis monkeys are helpless. Their mothers must carry the babies until they can walk. They'll **nurse** them for about seven months too. Males don't help with baby care. But they do protect their groups. Mothers care for their young for a year. Then the males leave to join other groups. The females stay with their mothers.

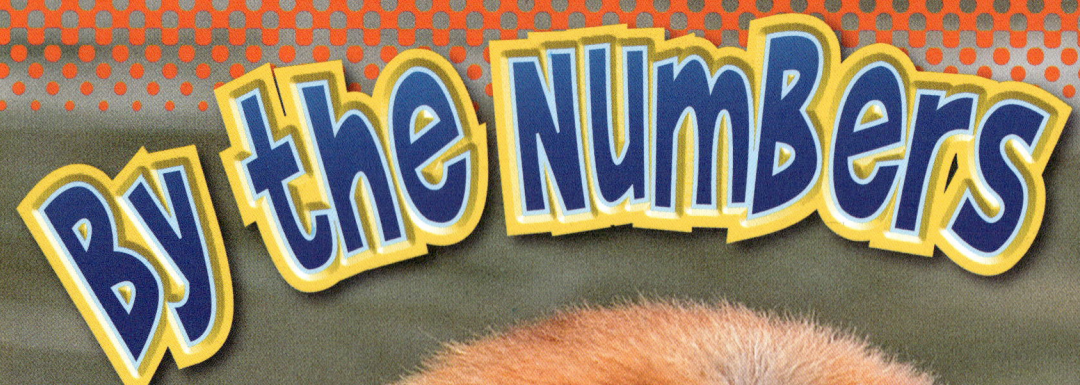

By the Numbers

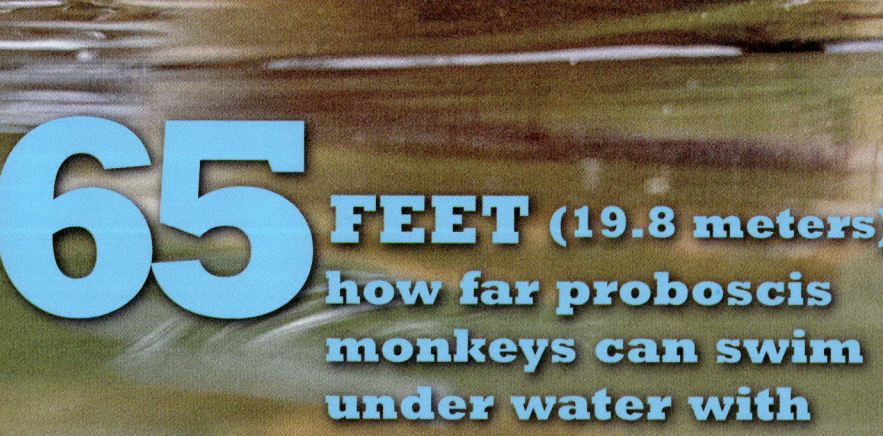

65 FEET (19.8 meters) how far proboscis monkeys can swim under water with one breath

20 to 30 INCHES (51 to 76 cm)

TAIL LENGTH

less than 20,000 estimated number left in the wild

5 YEARS OLD

AGE FEMALES CAN HAVE BABIES

15 to 20 YEARS LIFE SPAN

CHAPTER 4

These monkeys have many predators. In trees, they face danger from leopards and pythons. When a predator comes close, the monkeys might leap into water. But the water isn't always safe. That's where crocodiles hunt. But the monkeys are careful. They swim quickly to escape being eaten.

Proboscis Monkey Food Chain

This food chain shows what eats proboscis monkeys. It also shows what proboscis monkeys eat.

LEOPARDS

PYTHONS

CROCODILES

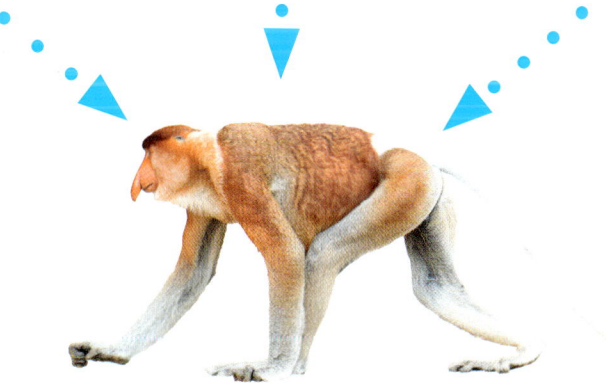
PROBOSCIS MONKEYS

INSECTS **LEAVES** **UNRIPE FRUIT**

Human Threats

Humans are also a threat to proboscis monkeys. They hunt the monkeys for food. Humans hurt the monkeys the most by causing habitat loss, though. People clear land for homes and farms. The monkeys have less land to live on. They need to leave trees more often. They must travel farther to find food. Proboscis monkeys are now **endangered**. If nothing is done, these creatures will disappear forever.

Protecting Proboscis Monkeys

People are trying to protect proboscis monkeys. Protecting these primates is a hard job, though. It is against the law to hunt the monkeys. But the monkeys are still in danger. More work is needed to protect these curious creatures.

The Proboscis Monkey Conservation Program researches the monkeys. It studies what they eat and where they live. Researchers hope to learn the best ways to help the primates.

29

GLOSSARY

attract (uh-TRAKT)—to draw someone or something to oneself

bacteria (bak-TEER-ee-uh)—a small living thing

digest (dy-JEST)—to change the food eaten into a form that can be used by the body

endangered (en-DAYN-jurd)—close to becoming extinct

mate (MAYT)—to join together to produce young

nurse (NURSE)—to feed a baby or young animal milk from the mother's body

predator (PRED-uh-tuhr)—an animal that eats other animals

social (SO-shul)—liking to be with and communicate with others

LEARN MORE

BOOKS

Bodden, Valerie. *Monkeys.* Amazing Animals. Mankato, MN: Creative Education, 2020.

Levy, Janey. *The Frightful Proboscis Monkey.* Nature's Freak Show: Ugly Beasts. New York: Gareth Stevens Publishing, 2020.

Rossiter, Brienna. *Saving Earth's Animals.* Saving Our Planet. Lake Elmo, MN: Focus Readers, 2022.

WEBSITES

Old World Monkeys
www.dkfindout.com/us/animals-and-nature/primates/old-world-monkeys/

Proboscis Monkey
www.britannica.com/animal/proboscis-monkey

Proboscis Monkey
www.nationalgeographic.com/animals/mammals/p/proboscis-monkey/

INDEX

B
babies, 4, 19, 20, 23

E
eating, 10, 20, 25, 27

F
features, 4, 7, 8–9, 10, 13, 19, 23

H
habitats, 4, 12–13, 17, 19, 24, 27

L
life span, 23

R
ranges, 13, 15

S
sizes, 6–7, 19

T
threats, 4, 16, 24, 25, 27, 28